INTEGRATING MATH IN THE REAL WORLD

THE MATH OF HOMES
AND OTHER BUILDINGS

Hope Martin and Susan Guengerich

J. WESTON
WALCH
PUBLISHER

Portland, Maine

User's Guide
to
Walch Reproducible Books

As part of our general effort to provide educational materials that are as practical and economical as possible, we have designated this publication a "reproducible book." The designation means that purchase of the book includes purchase of the right to limited reproduction of all pages on which this symbol appears:

Here is the basic Walch policy: We grant to individual purchasers of this book the right to make sufficient copies of reproducible pages for use by all students of a single teacher. This permission is limited to a single teacher and does not apply to entire schools or school systems, so institutions purchasing the book should pass the permission on to a single teacher. Copying of the book or its parts for resale is prohibited.

Any questions regarding this policy or requests to purchase further reproduction rights should be addressed to:

Permissions Editor
J. Weston Walch, Publisher
321 Valley Street • P. O. Box 658
Portland, Maine 04104-0658

1   2   3   4   5   6   7   8   9   10
ISBN 0-8251-3860-4

Contents

Introduction

In 1989, the National Council of Teachers of Mathematics (NCTM) developed the *Curriculum and Evaluation Standards for School Mathematics* to help teachers take their mathematics classes into the twenty-first century. The document calls for a curriculum that will help students solve problems and make connections between mathematics and other curricular areas, such as science, social studies, language arts, consumer education, and art. For math to be relevant, students must see how it relates to their lives outside of the mathematics classroom.

The lessons, activities, and projects in *Integrating Math in the Real World* have been designed to help students see interrelationships among subjects. Many of the activities are open-ended and encourage students to:

- work collaboratively to develop problem-solving strategies
- make connections between their life experiences and the math classroom
- develop self-confidence in their abilities to solve math problems

The Math of Homes and Other Buildings, one of the books in the *Integrating Math in the Real World* series, focuses on our homes, elements of construction and design, financial aspects of home ownership, and the tallest of our structures—skyscrapers. The lessons range from one-day activities that reinforce important skills and concepts to weeklong projects that involve students in multidisciplinary activities.

Each of the lessons is introduced with a Teacher Page. These Teacher Pages include the following sections:

Areas of Study

This section lists the mathematics skills of the lesson. Many of the activities are multifaceted and make use of a variety of math skills.

Concepts

This section contains a concise list of tasks for which the student will be accountable. If a rubric or grading matrix is being used, this list will be invaluable in developing specific criteria for assessment.

Materials

This section contains a list of materials that each student will need for the lesson unless otherwise specified. Collecting these items in advance will assure a smoother flow to the lesson.

Procedures

This section gives a brief description of the lesson with suggestions for the teacher. It is not meant to be a step-by-step recipe but merely a guide to help organize the lesson.

Assessment

Suggestions made are for effective, nontraditional ways to evaluate student achievement. It is suggested that student products be examined and that students be observed during the activity and questioned about their progress. Each lesson has one or more recommendations for journal questions that require a more in-depth understanding of the lesson concepts.

Extensions

Often some students need a more advanced or extended lesson. The suggestions discussed in this section can be used with a select group of students or with the entire class (if the lesson has been motivating and successful).

The lessons in *Integrating Math in the Real World: The Math of Homes and Other Buildings* have been designed to be teacher- and student-friendly. We have included lessons that make use of both metric and English measurements. This is not an accident. Since our students must learn both measurement systems, we have supplied activities that reinforce these diverse skills and concepts. In many cases, these lessons can be substituted for more traditional lessons found in mathematical texts.

Observation of Students

When students are active and working together, it is essential that the teacher walk around the room to become aware of the progress of the student groups and any problems that might arise. During these times it is possible to assess student understanding in a more formal way. While not every student can be observed each time, it is possible to perform a formal-type assessment at least twice during each grading period for each student. These observations can be shared with both parents and students during parent-teacher conferences.

A form, such as the one below, can be used to make the observations more consistent and simplify the process.

Name of Student _____

Criteria	4	3	2	1
How actively are students participating in group project?				
How well does student appear to understand concept of lesson?				
Is student actively listening to other members of group?				
Is student assuming positive leadership or problem-solving role?				
Comments:				

Using a Rubric for Performance Assessment

Authentic assessment is based upon the performance of the student and should be closely tied to the objectives of the lesson or activity. A rubric can be used to quantify the quality of the work. If the rubric is explained before the activity or project, students become aware of the requirements of the lesson. A grading matrix should be developed in which each of the objectives is examined using a five-point scale.

5 Student shows mastery and extends the concepts of the activity in new and unique ways

4 Student shows mastery of the concepts of the lesson

3 Student shows understanding, but there is a flaw in the presentation or reasoning

2 Student shows some understanding and has attempted completion, but there are some serious flaws in the presentation or reasoning

1 Student makes an attempt, but exhibits no understanding

0 Student makes no attempt

NCTM Standards Correlation

	Mathematics as Problem Solving	Mathematics as Communication	Mathematics as Reasoning	Mathematical Connections	Numbers and Number Relationships	Number Systems and Number Theory	Computation and Estimation	Patterns and Functions	Algebra	Statistics and Probability	Geometry	Measurement
Our New House	●	●	●	●	●	●	●			●	●	●
The 25 Tallest Buildings	●	●	●	●	●		●	●		●	●	●
My Front Door	●	●	●	●	●		●			●	●	●
The Math of Nails	●	●	●	●	●	●	●					
The Math of Rafters	●	●	●	●	●		●		●		●	●
The Math of Swimming Pools	●	●	●	●	●	●	●				●	●
The Math of Windows	●	●	●	●	●		●				●	●
Column Construction	●	●	●	●			●			●	●	●
Problem Solving	●	●	●	●	●	●	●	●	●	●		
Adjectives and Home Selection	●	●	●	●	●		●			●		●
Birdhouse Nets	●	●	●	●			●				●	●
Monthly Mortgage Payments	●	●	●	●	●		●					
Monthly Mortgage Payments and Interest Rates	●	●	●	●	●		●					
$25,000 Decorate-Your-Dream-Bedroom Contest	●	●	●	●	●	●	●				●	●
Math and Poetry				●								
Buying a House You Can Afford	●	●	●	●	●	●	●			●		
Housing Opportunity Index	●	●	●	●	●	●	●					

Our New House

Areas of Study

Computation, area and perimeter, percentages, minimum and maximum, reading charts and tables, scale drawing

Concepts

Students will:

- design a house based on cost and amenities chosen

- complete a Contractor Order Form within budget constraints

- draw a scale model of their house plans

- draw a scale model of their yard and landscape plans within zoning code regulations

- write and design an advertisement to sell their house

- calculate the selling price of their house

Materials

- one set of Our New House handouts per group

- one Cost of Your New Home slip per group

- ruler

- calculator

- examples of real estate ads

- colored pencils and/or markers

Procedures

Divide class into groups of three or four. Distribute packets and discuss the details involved in building a house. Things to consider include costs, zoning regulations, and personal taste. There's one rule: Each home *must* contain a kitchen and at least one bathroom.

When you are sure the students understand the concepts behind this project, distribute one cost slip to each group. Each group should receive a

different cost for their home so that there will be a variety of solutions to the problem.

The Price List contains a variety of options depending on the amount of money and personal taste of the group. All groups must fill out a Contractor Order Form and indicate their building costs. Their order should conform to building codes and must not exceed their budget.

Groups must submit Design Plans for their house (drawn to scale) and Yard and Landscape Plans (again drawn to scale). Finally, due to unforeseen circumstances, groups must sell their dream houses. Have them research real estate in your area and design an ad for the local paper. Tell them to add 6 percent to the asking price in order to pay the real estate agent's commission.

Assessment

1. Student products:
   - completed forms
   - scale drawings
   - real estate ad

2. Observation of students

3. Journal question(s):

   (a) Discuss the biggest problem your group had while deciding on options for your new home.

   (b) Why do you think communities have zoning regulations, and how did the ones in this project affect your decisions?

Extensions

- Research building costs and zoning regulations in your community. How do they compare with the ones in this project?

- Attend a zoning board meeting and interview members regarding the process they use in making their decisions.

Cost of Your New Home Slips

Hand out one price slip to each group, or place all the slips in a hat and have someone from each group draw a slip.

$50,000	$55,000	$60,000
$65,000	$70,000	$75,000
$80,000	$85,000	$90,000
$95,000	$100,000	$105,000
$110,000	$120,000	$125,000
$130,000	$135,000	$140,000
$145,000	$150,000	$155,000
$160,000	$165,000	$170,000
$175,000	$180,000	$185,000
$190,000	$195,000	$200,000
$250,000	$300,000	$500,000

The Math of Homes and Other Buildings

Our New House

Your group must design a home for all of you to live in together. Your home must contain a bathroom and a kitchen. The group will choose other amenities. You must follow the zoning regulations below. Read them very carefully.

Your basic construction cost is $75 per square foot. The costs of various amenities are on the Price List. After you have designed your home and decided upon the extras, complete the Contractor Order Form. You may not exceed your budget.

Zoning Regulations

1. All front yards must have a minimum 25-foot setback from the street. All backyards must be a minimum of 25 feet from the alley.

2. All homes, garages, sheds, and swimming pools must be 7 feet from the side lot lines.

3. Housing units for students must include separate bedrooms for each sex.

4. Swimming pools must be 4 feet from any building and must be enclosed with a 6-foot fence around the pool or the surrounding yard.

5. Only 50 percent of the lot may be covered with buildings.

6. The house can have, at most, three stories.

7. All houses must be 900 square feet, at minimum.

8. No on-street overnight parking is allowed. Parking places or a garage must be on the side or rear of the lot.

9. All zoning variances cost $100 per square foot and are added to the cost of the house.

3 *The Math of Homes and Other Buildings*

Name _____   Date _____

Our New House

Price List

Amenity	Description	Price
Building lot	66 by 150 ft.	$ 5,000
Building lot	132 by 150 ft.	$10,000
Building lot	264 by 150 ft.	$20,000
Building lot	264 by 300 ft., lakefront with dock	$50,000
Bathroom	$75 per sq. ft., plus standard fixtures	$ 5,000
Big bathroom	$75 per sq. ft., plus deluxe fixtures with whirlpool bath	$10,000
Whirlpool bath	10 ft. diameter	$ 5,000
Kitchen	$75 per sq. ft., plus standard fixtures	$10,000
Big kitchen	$75 per sq. ft., plus breakfast nook, microwave, preparation island, counter TV, deluxe appliances	$20,000
Fireplace	wood or gas log	$ 2,000
Sunroom	$75 per sq. ft., plus all windows, furniture, and fans	$ 5,000
Garage	one stall, attached to home	$ 5,000
Garage	two stall, attached or unattached	$10,000
Garage	three stall, with workshop area	$20,000
Swimming pool, aboveground	outdoor 20 by 40 ft.	$10,000
Swimming pool, in-ground	$75 per sq. ft., plus equipment	$30,000
Deck	$75 per sq. ft., plus materials	$ 1,000
Game room	$75 per sq. ft., plus equipment	$ 2,000
Fitness room	$75 per sq. ft., plus equipment	$ 8,000
Computer room	$75 per sq. ft., plus equipment	$10,000
TV room	$75 per sq. ft., plus equipment	$ 4,000
Home security		$10,000
Sound system for whole house		$ 3,000
Fence	$10 per ft.	

         The Math of Homes and Other Buildings

Our New House

Contractor Order Form

Item	Basis for Calculation	Total Cost
Lot	Size	
Total area of house	Square feet by $75	
Bathroom	Standard or deluxe	
Kitchen	Standard or deluxe	

Our New House

Design Plans

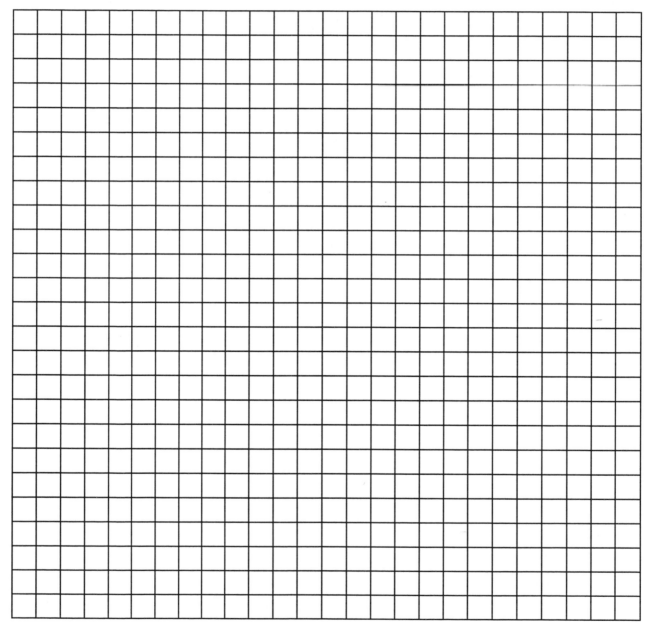

Our New House

Yard and Landscape Plans

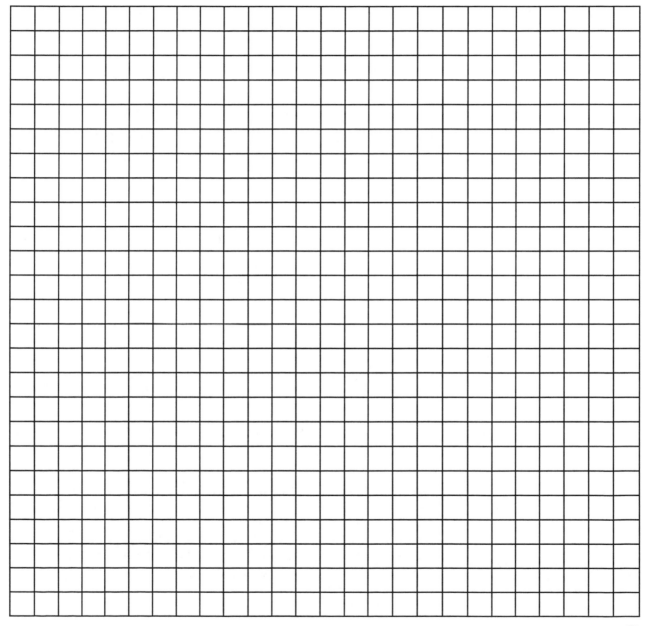

Our New House

For Sale

Due to unforeseen circumstances, your group must sell its new home.

Using newspapers, trade papers, or real estate booklets, study the real estate advertisements in your community. Design an ad to sell your new home. List its features, using lots of adjectives. Include a drawing of your home. Be sure to give the listing price. Remember, the real estate agent will earn a commission of 6 percent of the selling price.

The 25 Tallest Buildings in the World

Areas of Study

Reading charts and tables, statistics, analysis of data, percentages, box-and-whisker plots, scale drawing, scatterplot

Concepts

Students will

- examine the table to find the city with the most skyscrapers, calculate the percentage of the world's skyscrapers in that city, and research construction techniques of tall buildings

- design a box-and-whisker plot of the 25 tallest buildings

- design a scatterplot of the relationship between the height of the building and the number of stories

- draw three of the buildings to scale

- indicate selected skyscrapers on a map of the world

Materials

- The 25 Tallest Buildings in the World handouts

- calculator

- pictures of buildings (if available)

Procedures

Give each student a copy of the table. Discuss the new criteria set by the Council on High Buildings for judging the heights of buildings and why the criteria were changed. The list in the table is based on previous criteria for the tallest structures, which included all towers and extensions.

Allow each student to work with a partner to answer questions and then, as a group, develop the box-and-whisker plot. Make an overhead transparency of the page and work through the steps with students.

The scatterplot is a graph of the relationship between the number of stories in the building and its height.

The scale drawings will be more interesting if students see pictures of the buildings. A good book to use with this project is: *Skyscrapers,* by Judith Dupré, Black Dog & Leventhal Publishers.

Assessment

1. Student products:
   - successful completion of worksheet
   - box-and-whisker plot
   - scatterplot
   - scale drawing

2. Observation of students

3. Journal question(s):

   (a) If a building had 100 stories, about how tall do you think it would be? Explain your answer.

   (b) What do you think the building of the future with a height of 2,000 feet will look like? How would you describe the type of construction needed to build it?

Extensions

- Enter data into a graphing calculator and make box-and-whisker plots and scatterplots with the calculator.

- Use a graphics calculator with higher level math students to develop a line of best fit.

The 25 Tallest Buildings in the World

Below is a list of the 25 tallest buildings in the world in 1995, according to the Council on High Buildings and Urban Habitat, Lehigh University. The building listed as UC (under construction) may now be completed. Search the Web to see if there are any taller buildings now.

Use the information in the table to answer some questions about skyscrapers.

Rank	Building	City	Year Built	Stories	Height in Feet
1	Petronas Towers 1 & 2	Kuala Lumpur	1996	88	1,476
2	Sears Tower	Chicago	1974	110	1,454
3	Jin Mao Building	Shanghai	1998 UC	88	1,379
4	World Trade Center North	New York	1972	110	1,368
5	World Trade Center South	New York	1973	110	1,362
6	Empire State Building	New York	1931	102	1,250
7	Central Plaza	Hong Kong	1992	78	1,227
8	Bank of China Tower	Hong Kong	1989	70	1,209
9	Tuntex & Chien-Tai Tower	Kaohsiung	1997	85	1,140
10	Amoco	Chicago	1973	80	1,136
11	John Hancock Center	Chicago	1969	100	1,127
12	Sky Central Plaza	Guangzhou	1996	80	1,056
13	Baiyoke Tower II	Bangkok	1997	90	1,050
14	Chrysler Building	New York	1930	77	1,046
15	Shenzhen Avic Plaza	Shenzhen	1997	63	1,025
16	NationsBank Plaza	Atlanta	1992	55	1,023
17	1st Interstate Trade Center	Los Angeles	1989	75	1,018
18	Texas Commerce Tower	Houston	1982	75	1,000
19	Ryugyong Hotel	Pyongyang	1995	105	984
20	Two Prudential Plaza	Chicago	1990	64	978
21	1st Interstate Bank Plaza	Houston	1983	71	972
22	Landmark Tower	Yokohama	1993	70	971
23	311 South Wacker Drive	Chicago	1990	65	959
24	Jubilee St./Queen's Rd.	Hong Kong	1997	69	958
25	1st Canadian Place	Toronto	1975	72	952

Name _____   Date _____

The 25 Tallest Buildings in the World

Find the height of each story in the top 10 tallest buildings. Write your answers in the space provided:

Building	Height	Number of Stories	Height of Each Story
Petronis Towers			
Sears Tower			
Jin Mao Building			
World Trade Center N			
World Trade Center S			
Empire State Building			
Central Plaza			
Bank of China Tower			
Tuntex & Chien-Tai			
Amoco			

The average height of one story for these 10 buildings is _____

Which skyscraper held the record of tallest building in the world for the longest period of time? _____

Why do you think it took so long for a taller building to be built? Do some research about construction of tall buildings to help you answer this question.

_____

_____

_____

Which city on this list has the largest percentage of skyscrapers? _____

What is the percentage?_____

Use the map of the world on page 15 to locate each of the 25 tallest buildings in the world. Use any reference materials you need to be as accurate as possible.

The 25 Tallest Buildings in the World

Use the graph to plot heights of the buildings and the number of stories. What type of relationship do you see?

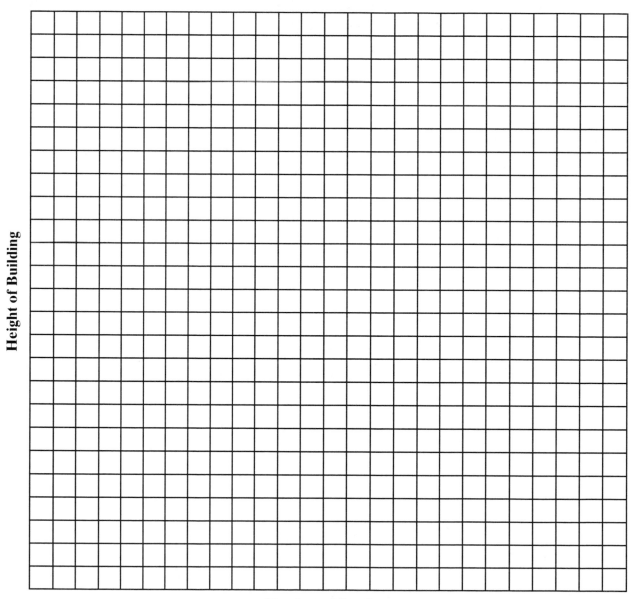

Height of Building

Number of Stories

The 25 Tallest Buildings in the World

Using the data from the table, design a box-and-whisker plot to help us analyze this information. Follow these steps:

1. Find the median (or middle) of the data. Write it here: _____

    You now have 50 percent of the heights below this point and 50 percent above this height.

2. Find the median of the lower half (lower quartile—LQ). Write it here: _____

3. Find the median of the upper half (upper quartile—UQ). Write it here: _____

4. Label the number line below with the appropriate coordinates to accommodate the tallest and shortest buildings on the list. (The tallest is 1,476 feet and the shortest is 952 feet.) You can do that by answering these questions:

    a. What is the range of data? _____

    b. How many coordinates are there on the line? _____

    c. What units will work? _____ Label the line with these units.

5. Put a dot on the coordinate representing the shortest building on the list and a dot on the coordinate representing the tallest building on the list. Connect these points with a line—this is the whisker!

6. Put a vertical line on the median coordinate, on the UQ, and on the LQ. When you connect these, you have formed a rectangular box.

Heights of Buildings

What percentage of the buildings are in each of the four sections of this graph?

_____

Explain how you know this. _____

_____

The 25 Tallest Buildings in the World

Use this grid to draw any 3 of the 25 tallest buildings to scale. Use reference materials to help you make them look realistic.

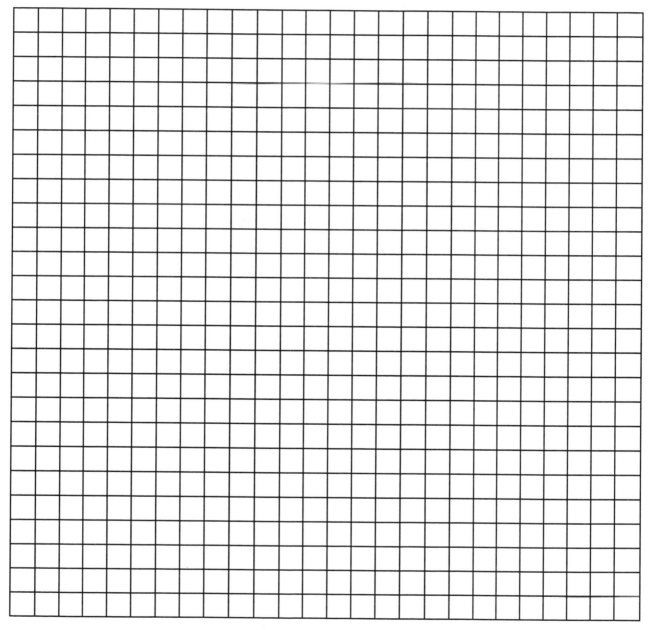

The 25 Tallest Buildings in the World

On the map below, locate as many skyscrapers as you can. Use reference materials to help with accuracy. Since many of the buildings are located in the same areas of the world, you may want to indicate only the one or two tallest buildings in that city.

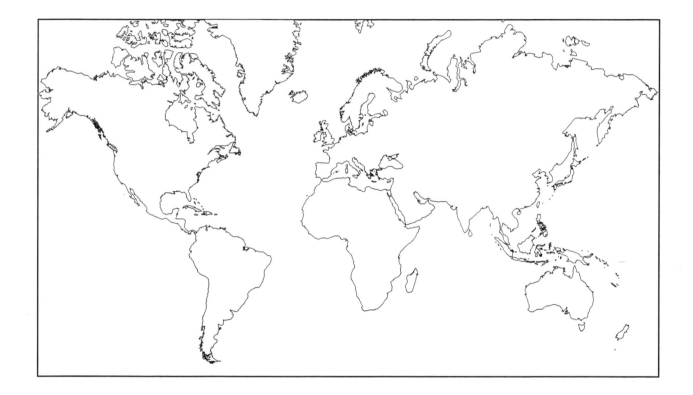

My Front Door

Areas of Study

Measurement, data collection, statistics, sorting data from least to greatest, stem-and-leaf plot

Concepts

Students will:

- measure the width of their front door
- collect class data of their measurements
- sort their data from least to greatest
- make a stem-and-leaf plot of the data
- find the range, mode, median, and mean of the data

Materials

- My Front Door handout
- overhead transparency of Class Data Collection Sheet
- overhead transparency of stem-and-leaf plot
- calculators

Procedures

Give students a copy of the handout, My Front Door, to take home and record their measurements on. Emphasize that all measurements should be to the nearest inch.

The next day, students record their data on the Class Data sheet. Then have students transfer their data to the stem-and-leaf plot. It is easier to rewrite the *leaves* and sort the numbers from least to greatest. Transfer this organized data back to your Class Data sheet.

Students use this organized data to find the range, mode, median, and mean.

Assessment

1. Student products
2. Observation of students
3. Journal question:
   - Congress passed a law called *The Americans with Disabilities Act* to make sure that people with disabilities can lead normal lives. The law says that doors should be 36 inches wide to accommodate people with wheelchairs and walkers. What percentage of our doors would meet this requirement?

Extensions

- Students can be asked to find the perimeter and surface area of their door.
- Have students design a door that might reflect their interest and tastes 25 years from now.

My Front Door

The standard height for all doors is 80 inches, or 6 feet 8 inches. Most people can enter this door without fear of bumping their head. However, the width of front doors will vary. Some are very narrow and others are quite wide. How wide is the entrance to your residence? Measure the width of your front door and write that measurement here (in inches). _____

Entrances have characteristics that can give clues to what lies beyond the door. Some doors are warm and friendly, some are humorous, some give warning, and some are quite plain and anonymous. In the space below, sketch the entrance to your residence.

List some adjectives that you feel would describe your door:

My Front Door

Class Data Collection Sheet

Name	Data (Door Width)	Data Sorted from Least to Greatest

Using the data collected from your class, determine the range, mode, median, and mean.

Range

Mode

Median

Mean

My Front Door

Let's transfer the data from our Class Data Collection Sheet to a stem-and-leaf plot.

| 3 | 5   means 35 inches

```
0 |
1 |
2 |
3 |
4 |
5 |
6 |
7 |
8 |
9 |
```

Now let's rewrite each stem in numerical order:

```
0 |
1 |
2 |
3 |
4 |
5 |
6 |
7 |
8 |
9 |
```

The Mathematics of Nails

Areas of Study

Computation, fractions, rounding up, and reading tables

Concepts

Students will:

- calculate the length of a nail that protrudes after being nailed into a common 2" × 4" piece of lumber

- calculate the number of pounds of nails that must be purchased to have 1,000 nails; the answers are rounded up to the nearest quarter pound

Materials

- The Mathematics of Nails handout

- calculator

- various nails and piece of 2" × 4" board for demonstration

Procedures

Discuss the various sizes and types of nails. Different sized nails are used for different projects. Nails are classified by size using an old system of pennies. A 2" × 4" board is actually $1\frac{1}{2}$" × $3\frac{1}{2}$" when sold. Students will calculate how much of the nail will protrude from the 2" × 4" by subtracting $1\frac{1}{2}$" from the length of each nail. Students then calculate the number of pounds of nails that must be purchased to have 1,000 nails for a project. The answer is rounded up to the nearest quarter pound to ensure enough nails to finish a project and

because many hardware stores sell nails only in quarter-pound increments.

Solutions

Size	Protruding	Size	Pounds
8 d	1"	5 d	4
10 d	$1\frac{1}{2}$"	6 d	6
16 d	2"	8 d	10
20 d	$2\frac{1}{2}$"	10 d	$15\frac{1}{4}$
30 d	3"	16 d	$21\frac{1}{4}$
40 d	$3\frac{1}{2}$"	20 d	$33\frac{3}{4}$
60 d	$4\frac{1}{2}$"	40 d	58

Assessment

1. Student product:
   - completed handout

2. Observation of students

3. Journal question:
   - Why would a carpenter almost always round measurements up rather than to the nearest?

Extension

- Have students research this question: Why are nail sizes and strengths closely monitored by the government and construction companies?

The Mathematics of Nails

The length of common nails is measured in units called **pennies.** The abbreviation of penny is **d**. A 6 d nail is 2 inches long and a 10 d nail is 3 inches long.

Because a standard 2" × 4" shrinks from drying and is planed for smoothness, the board is actually $1\frac{1}{2}$" × $3\frac{1}{2}$" when sold. If a 10-penny nail were driven through a standard 2" × 4", $1\frac{1}{2}$" of the nail would stick out (protrude) from the board. 3" – $1\frac{1}{2}$" = $1\frac{1}{2}$". Complete the chart by calculating the length of the nail protruding.

Nails are sold by the pound. The larger the nails, the fewer per pound. Use the data below to calculate the number of pounds you need to have 1,000 nails. Round up to the nearest quarter pound to assure enough nails to complete a project.

Nail Size	Length in Inches	Inches Protruding
8 d	$2\frac{1}{2}$	
10 d	3	
16 d	$3\frac{1}{2}$	
20 d	4	
30 d	$4\frac{1}{2}$	
40 d	5	
60 d	6	

Nail Size	Nails per Pound	Pounds for 1,000 nails
5 d	254	—
6 d	167	
8 d	101	
10 d	66	
16 d	47.4	
20 d	29.7	
40 d	17.3	

The Mathematics of Rafters

Areas of Study

Using the Pythagorean Theorem, rounding

Concepts

Students will:

- use the Pythagorean theorem to calculate the length of roof rafters
- round their answers up to the nearest tenth

Materials

- The Mathematics of Rafters handout
- calculator

Procedures

Read with students the description of rafters. Discuss how they could use the Pythagorean theorem to find the length of the rafter if they were aware of the span and rise.

Allow students time to calculate the length of the rafters using their calculators. Have them round their answers up to the nearest tenth.

Solutions

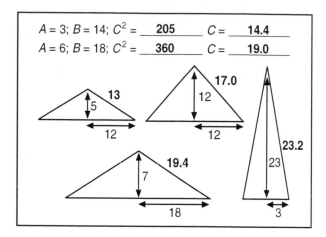

$A = 3$; $B = 14$; $C^2 = $ __205__ $C = $ __14.4__

$A = 6$; $B = 18$; $C^2 = $ __360__ $C = $ __19.0__

Assessment

1. Student products
2. Observation of students
3. Journal question:
   - Suggest a way we might find the rise of the roof if we knew the span and the length of a rafter.

Extension

- Have students compute the approximate length of the rafters on structures using roofing materials.

The Mathematics of Rafters

Rafters support the roof of a house. In most homes, the rafters are wood. The builder must cut the rafters to size or purchase prefabricated rafters. Builders must calculate the rafter length to construct a roof and to calculate the amount of roofing materials that will be needed. The Pythagorean theorem can be used to determine the rafter length. $A^2 + B^2 = C^2$ is used, where A is the rise, B is $\frac{1}{2}$ the span, and C is the length of the rafter.

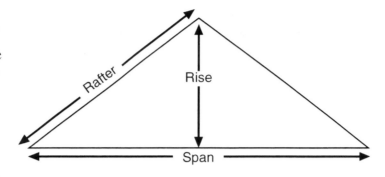

Use $A^2 + B^2 = C^2$ to find the value of C^2. Use your calculator and the square root $\sqrt{\ }$ function to find the value of C. Round C up to the nearest tenth.

$A = 3; B = 14; C^2 =$ _____ $C =$ _____

$A = 6; B = 18; C^2 =$ _____ $C =$ _____

Find the rafter length for the following roofs. The rise and span are given. Round your answers up to the nearest tenth.

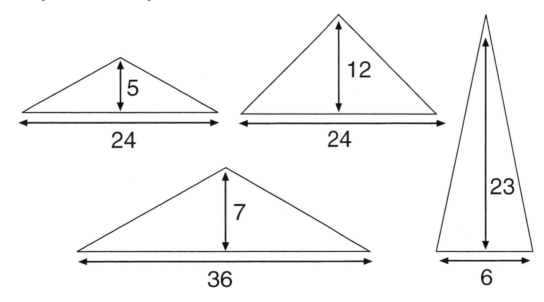

The Mathematics of Swimming Pools

Areas of Study

Computation, volume conversion of measurement units, and rounding

Concepts

Students will:

- calculate the volume of various shaped swimming pools
- convert cubic feet to gallons
- determine the cost to fill the pool
- round to the nearest gallon

Materials

- The Mathematics of Swimming Pools handout
- calculator

Procedures

Review the concepts of calculating volume with students. Discuss the volume of a swimming pool and the cost of filling it. Distribute handouts. Students may work in groups to find the volume in cubic feet and then convert the cubic feet to gallons. Round to the nearest gallon. Students will then calculate the price of filling the pool, using $5 for 1,000 gallons. Round to the nearest cent.

Solutions

Diving pool	**In-ground pool**	**Backyard pool**
Cubic feet = 24,000	Cubic feet = 11,700	(semicircular ends)
Gallons = 178,032	Gallons = 86,791	Cubic feet = 2,245
Cost = $890.16	Cost = $433.95	Gallons = 16,655
		Cost = $83.28

Backyard pool	**Hot tub**	
(rectangular)	Cubic feet = 314	
Cubic feet = 3,200	Gallons = 2,329	
Gallons = 23,738	Cost = $11.65	
Cost = $118.69		

Assessment

1. Student product:
   - completed handout
2. Observation of student
3. Journal questions:
   (a) What are some of the "other" expenses of owning a pool?
   (b) Many communities require fences to surround all pools. How many feet of fence would be required to surround the top of the backyard pool with the semicircular ends?

Extensions

- Research the cost of water in your community and recalculate the cost of filling the pools.
- Research the size of a local public pool; then calculate its volume and the cost to fill it with water from your community.

The Mathematics of Swimming Pools

Filling a swimming pool takes lots of water. Just how much water? Use the value 3.14 for π and calculate the volume of these swimming pools in cubic feet. Convert cubic feet to gallons and round to the nearest whole gallon. If water costs $5 per 1,000 gallons, how much would it cost to fill the swimming pools?

1 cubic foot = 7.418 gallons

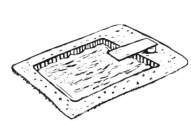

Diving Pool

Width 40 feet
Length 40 feet
Depth 15 feet
Cubic feet = _____
Gallons = _____
Cost to fill = _____

Hot Tub

Diameter 10 feet
Depth 4 feet
Cubic feet = _____
Gallons = _____
Cost to fill = _____

Backyard Pool (rectangular)

Width 20 feet
Length 40 feet
Depth 4 feet
Cubic feet = _____
Gallons = _____
Cost to fill = _____

Backyard Pool (semicircular ends)

Width 12 feet
Length 40 feet
Depth 5 feet
Cubic feet = _____
Gallons = _____
Cost to fill = _____

In-Ground Pool

Width 30 feet
Length 60 feet
Depth 4 feet to 9 feet
Cubic feet = _____
Gallons = _____
Cost to fill = _____

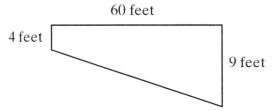

60 feet

4 feet

9 feet

Use area of trapezoid multiplied by the width of the pool to find the volume in cubic feet.

The Math of Homes and Other Buildings

The Mathematics of Windows

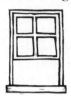

Areas of Study

Area, perimeter, converting units of measurement, and reading charts

Concepts

Students will:

- read window codes and convert the codes to feet and inches

- convert feet and inches to total inches

- calculate the perimeter and area

Materials

- The Mathematics of Windows handout
- calculator

Procedures

Give each student a copy of the handout and discuss the variety of windows found in buildings. Discuss some problems that builders and contractors might have when purchasing windows. Explain the code used to classify a window's width and height. Use the example on the student page to show the calculations needed to complete the handout.

Solution

Window Code	Window Width	Window Height	Width in Inches	Length in Inches	Perimeter in Inches	Area in Inches
2846	2'8"	4'6"	32	54	172	1,728
3444	3'4"	4'4"	40	52	184	2,080
3852	3'8"	5'2"	44	62	212	2,728
4468	4'4"	6'8"	52	80	264	4,160
2646	2'6"	4'6"	30	54	168	1,620
3268	3'2"	6'8"	38	80	236	3,040
32410	3'2"	4'10"	38	58	192	2,204
8068	8'0"	6'8"	96	80	352	7,680

Assessment

1. Student product:
   - completed handout

2. Observation of students

3. Journal questions:

   (a) Compare a 3268 window to a 6832 window. What are the perimeters and areas?

   (b) Why would a heating and air conditioning installer need to know the size and number of windows?

   (c) The standard door is 6'8", or 80 inches. List basketball players or other famous people who would have trouble entering a standard door.

Extensions

- Have students measure and write the window code for windows found in the classroom or in their homes.

- Measure windows in students' homes and find the mean, mode, median, and range for the widths and heights.

- A window has a perimeter of 188 inches and an area of 2,040 square inches. What are the dimensions of the window? Write the window code for this window.

Name _____ Date _____

The Mathematics of Windows

What size are the windows of your classroom? Are the classroom windows the same size as the windows in your home? Are all the windows where you live the same size? Windows are manufactured in many sizes and styles. A builder must be able to communicate the size windows needed. A classification system has been developed to help builders and window suppliers.

Windows are assigned a code number that corresponds to the size of a window. The code gives the width of the window and then the height. A window might be 2846. This is read as 2 feet 8 inches as the width of the window and 4 feet 6 inches as the height of the window. The 2846 window is called a two eight, four six. (Do not confuse 2846 to mean 28 inches by 46 inches.)

In the chart below, determine the width and height of a window in feet and inches. Change feet and inches to inches and calculate the area and the perimeter of the window.

Window Code	Window Width	Window Height	Width in Inches	Length in Inches	Perimeter in Inches	Area in Inches
2846	2'8"	4'6"	32	54	172	1,728
3444						
3852						
4468						
2646						
3268						
32410						
8068						

Column Construction

Areas of Study

Problem solving, data collection, organization and analysis, statistics, line plot, scientific method, measurement

Concepts

Students will:

- design a column, following specific construction guidelines

- use the experimental method to ascertain the column's strength

- collect data from the class

- analyze why some designs are stronger than others

- use a line plot to organize data for further analysis

- find the range, median, mode, and mean of the data

Materials

- paper, $8\frac{1}{2}$" × 11"

- clear tape or glue

- a collection of about 30 paperback books (all must be the same size)

- overhead transparencies of Data Collection and Data Analysis sheets

- Guidelines handout

- ruler

Procedures

Give each student a copy of Column Construction Guidelines. Review guidelines to be sure students understand them. Each student is respon-sible for designing a column, carefully following these guidelines:

1. use only one sheet of $8\frac{1}{2}$" × 11" paper

2. use only 25 cm of tape or 1 ml of glue

3. ensure the column holds the books at least 15 cm above the tabletop for at least 5 seconds.

Student will test their final column design in front of the class. On the overhead transparency, record the results on the Data Collection sheet. Be sure to examine the column to make sure that it follows the guidelines exactly.

Once all of the data are collected, use the line plot on the Data Analysis sheet to organize it. The coordinates will be determined by your class's final data. Have students find the statistics asked for on the sheet.

Follow-up discussion should focus on the best designs and on what changes could have been made to improve the designs.

Assessment

1. Student product:
   - column

2. Observation of students

3. Journal question:
   - Describe the development of your column designs from the first to the one you used in the final competition.

Extension

- Using an $8\frac{1}{2}$" × 11" piece of corrugated card-board, conduct the same experiment. Then analyze the results of using this type of paper.

Column Construction

Guidelines

1. You may use only one sheet of $8\frac{1}{2}$" × 11" plain white paper and 25 cm of tape or 1 ml of glue.

2. You will use paperback books to determine the strength of your column. All the books must be the same size.

3. The column must hold the paperback books at least 15 cm above the table.

4. After each book is added, you must wait 5 seconds before adding the next book.

5. Count the number of books your column held. Do not count the book that caused the collapse.

6. Report your data to the class.

7. Use a line plot to organize the data from the entire class.

8. Record, sort, and analyze the data collected from your class. Determine the range, mean, median, and mode of the data.

(continued)

         The Math of Homes and Other Buildings

Name _____ Date _____

Column Construction

Guidelines *(continued)*

Column Construction Data Sheet

Name	Number of Books Your Column Held	Design Elements of Your Column

Column Construction

Data Analysis

Use this line plot to help organize the data from the class data table.
Use the data your class collected to label the coordinates.

├─┼─┼─┼─┼─┼─┼─┼─┼─┼─┼─┼─┼─┼─┼─┼─┼─┼─┼─┼─┤

The range of the data is: _____

The mode of the data is: _____

The median of the data is: _____

The mean of the data is: _____

Discuss with your class the engineering design that was the strongest and held the most books.

How do you think the design could be improved following the same construction guidelines? Write your answer on the back of this sheet.

Problem Solving

Areas of Study

Problem solving, area, perimeter, percentages, logical reasoning, computation, factors, and number theory

Concepts

Students will:

- solve problems in groups and show the mathematical reasoning
- explain the group solution to the class
- use a variety of math skills to solve problems

Materials

- Problem Solving handouts
- calculator

Procedures

Review problem-solving strategies, such as making an organized list, working backward, and looking for a simpler solution. Distribute one problem to each group. Use normal classroom procedures for group work. Each student should write an explanation after the group has found a solution.

Solutions

Dimensions of the Basement

The Williamses will need 175 square yards or 1,575 square feet of floor covering. The 5 is in the units place. The product of the two sides of the basement must end in a 5, so the sides must have units digits of 5×5 or 5×1. The perimeter of 160 feet indicates that the sum of the two sides is 80 feet. Both addends must end in 5 to have a sum with units digits of 0. The sides of the basement are 45 by 35.

Goats and the Lawn

Both arrangements would result in the same amount of grass being eaten. It would most likely be more efficient to own just one goat. The yard is 60 by 60 with an area of 3,600 square feet. The circle would have a radius of 30 feet and an area of 2,826 square feet. The circle is 78.5% of the yard.

Home Improvement Purchase

The apartment maintenance manager is buying house numbers. The number 3 is one digit and would cost one dollar. Since 30 is two digits, it would cost $2. Both 300 and 350 are three-digit numbers and would cost $3.

Where Is This House?

The house is located at the North Pole.

Apartment Number?

The apartment number does not have a factor of 2, 3, 4, 5, or 6. It does have a factor of 11. The number will end in 1 because if a number is divided by 5 and the remainder is 1, the units digit is one or six. The number is odd, so the apartment number has a unit digit of 1. The apartment number is 781. It is the product of 11 and 71.

Assessment

While each problem has a correct solution, there are many different ways to reach it. Part of the lesson should include showing some of the many ways students reached their solutions. Different or unique strategies should be examined.

Extension

- Students can write their own problems about homes and housing and ask other members of the class to solve them.

Problem Solving

Dimensions of the Basement

The Williams family decided to convert their basement into a rec room. They knew they needed 175 square yards of floor covering and 160 feet of baseboard trim. What are the dimensions of the room?

Problem Solving

Goats and the Lawn

Mr. Green and his daughters hate to mow the lawn, so they decided to let goats eat the grass. They have a square backyard. Each side is 60 feet long. Would one goat chained to the center of the yard or four goats chained to each of the four corners be more efficient?

What percent of the yard would they still need to mow?

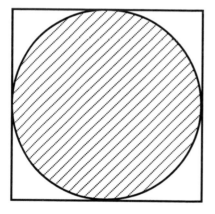

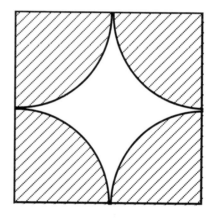

% of yard to mow: _____

% of yard to mow: _____

Problem Solving

Home Improvement Purchase

An apartment maintenance manager goes to a home improvement store. He finds that 3 will cost him $1; 30 will cost him $2; 300 will cost him $3. When he asked about the cost of 350, it was also $3. What was he buying?

Problem Solving

Where Is This House?

A builder is planning a house. She measures 50 feet to the south, 50 feet to the east, and then 50 feet north and is exactly where she started. Where is this house located? What is your reasoning?

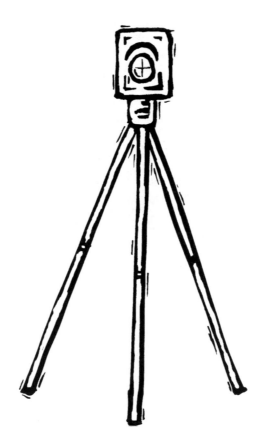

Problem Solving

Apartment Number?

When asked her apartment number, Miss Prime replied, "Figure it out yourself. If you divide my apartment number by 2, 3, 4, 5, or 6 there will be a remainder of 1. There is no remainder when you divide by 11. There is no lower apartment number that has the same characteristics." What is Miss Prime's apartment number?

Adjectives and Home Selection

Areas of Study

Newspaper research, ratio, percentages

Concepts

Students will:

- find advertisements of houses for sale in newspapers and other sources
- paste the ad on the handout
- find all the adjectives used in the ad
- calculate the ratio of adjectives to total words
- calculate the percent of adjectives in two different ads

Materials

- newspapers and other sources of home ads
- Adjectives and Home Selection handout
- scissors
- calculator
- glue

Procedures

Students will use home-buying guides in local newspapers and other sources to find a descriptive ad for a home. They will:

1. locate and list all the adjectives
2. count the total number of words
3. find the ratio of number of adjectives to total word count
4. calculate the percentage of adjectives in the ad

Students are asked in the last question of the handout to locate a different type of ad in the paper, to find the adjectives in that ad, and to calculate the percentage. Students should reflect on the similarities or differences they found and report them.

Assessment

1. Student products
2. Observation of students
3. Journal question:
   - included in lesson

Extension

- Ask students to find three substantially different homes, ranging from inexpensive to luxurious. Have students analyze whether the cost of the house affects (1) the length of the ad and (2) the ratio of adjectives to total words.

Adjectives and Home Selection

Use the home-buying guide in your newspaper or other sources to select a home to purchase. Cut out the ad and paste it in the space provided.

Paste
the advertisement
for
your
home
selection
here.

List all the adjectives used in the advertisement:

_____

_____

_____

How many adjectives are in the advertisement? _____

How many total words are in the advertisement? _____

What percentage of the words in the advertisement are adjectives? _____

Do you think the percentage of adjectives is the same as or different from ads in other parts of the newspaper? Why? Find an example that supports your idea.

Birdhouse Nets

Areas of Study

Two- and three-dimensional geometry, spatial visualization, problem solving, area, surface area, scale models

Concepts

Students will:

- discover hexominoes which form cubes when folded

- design rectangular prisms

- design a birdhouse using 1 cm graph paper

- convert their design using the scale 1 cm: 1 in

- build a birdhouse using foam board or poster board

- identify the shapes that comprise their birdhouse

- find the area of the faces and surface area of their birdhouse

Materials

- Birdhouse Nets handouts

- scissors

- glue gun (if working with foam board)

- clear adhesive tape (if working with poster board)

- poster board or foam board (half sheet for each student)

- markers or paint

- calculator

- utility knife (if working with foam board)

Note: Students will need adult supervision if working with utility knives and glue guns.

Procedures

Students will examine the sample hexominoe and learn that when it is cut on the solid lines and folded on the dotted lines it forms a cube. None of the three examples shown on the bottom of page 42 folds into a cube. Ask students to use the full sheet of graph paper to find other hexominoes which are cubic nets—and which do fold into cubes.

Next, have students examine a net for a rectangular prism. Ask them how this net differs from the cubic net. They may say that the sides are rectangles and not cubes. Make sure they understand the correlation between the sides of each face; that the width of the front of the box and the width of the top of the box must have the same measure. Working in groups, students will design their own rectangular prism. Cereal boxes can serve as models.

Finally, have students design their own birdhouses. Students need to understand that their birdhouse must have right angles at the base or their net will not fold correctly. Many commercial birdhouses have bases that are smaller than their tops; this is possible only if the sides are assembled individually—they cannot be made from a net. This is an example of a successful net:

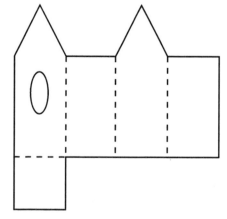

The roof is a rectangle that is folded in the middle and then glued on the top. Students should be encouraged to be original with their designs, but be sure that they cut their design out of graph paper before attempting to cut the poster board or foam board.

Students can decorate their birdhouses, using markers or paint. Either works well on foam board.

Assessment

1. Student products:
   - successful completion of birdhouse
   - quality of written responses
2. Observation of students
3. Journal question(s):

   (a) Design an open-top box using only 5 squares.

   (b) Draw a birdhouse net, measure its faces, and calculate its surface area.

Extension

- Have students compute the volume of their birdhouses. This will involve finding the volume of rectangular and triangular prisms. Once the birdhouse volumes have been estimated, students can fill them with rice and compare the volumes.

Birdhouse Nets

We are going to figure out how to design our own unique bird-house using a minimum number of pieces. We'll start out with a simple shape. Look at the geometric figure below. What do you think would happen if we cut along the solid lines and folded on the dotted lines? Use your graph paper to copy the shape, cut it out, and fold it as shown. Describe your shape: _____

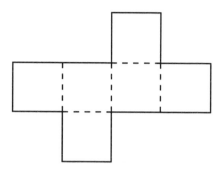

This is called a **hexominoe net** because it is a two-dimensional representation of a three-dimensional shape. If folded correctly, it will form a cube. Are there other nets that we can design that will give us a cube?

Look at these three hexominoes and decide which of them we will be able to fold into a cube. If you need to, cut them out to check your reasoning.

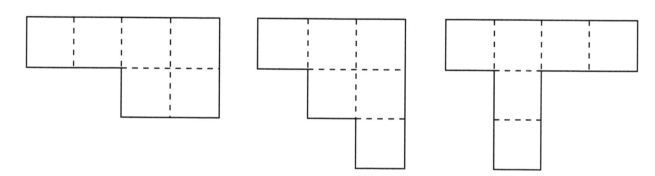

Use your graph paper to design as many different cubic nets as you can.

Birdhouse Nets

Use this graph paper to design hexominoes. There are 35 hexominoes but not all of them are cubic nets. Circle the ones that form cubes.

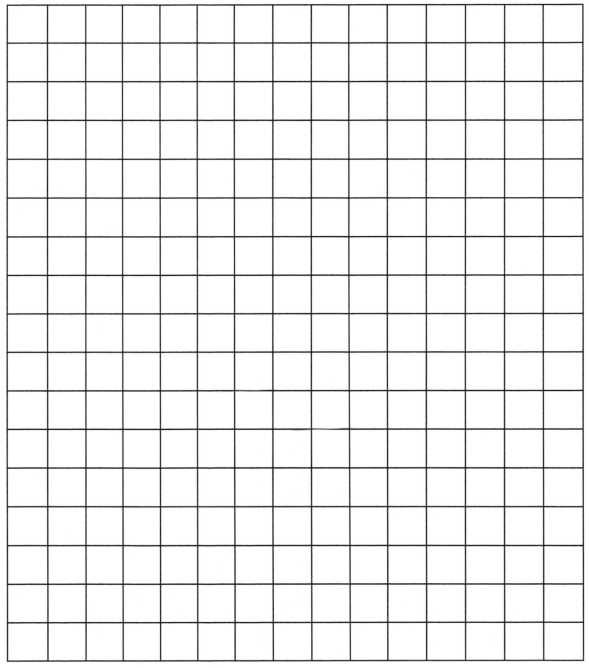

Birdhouse Nets

Look at the net below. If you cut it on the solid lines and fold it on the dotted lines, what three-dimensional shape do you think you will get? Why? Write your answer on the back of this paper.

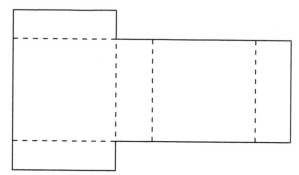

Use this graph paper to design a rectangular prism (the geometric name for the cereal-box shape shown above. See if you can put the top, bottom, and sides in different places and still end up with the same shape.

Birdhouse Nets

Now that we know what a net is, the fun begins! We are going to design nets that can be used to build **birdhouses**. We will follow these steps:

1. Design a birdhouse net using 1-cm grid paper. Each cm on the paper will represent 1 inch on your birdhouse. Use the graph paper below.

2. Examine available pictures of birdhouses to see the shapes of the front (and back) and the two sides.

3. Decide what type of roof you would like and design a roof that will fit over the birdhouse.

4. Cut out your net to be sure that it will work.

5. Use your piece of foam board to re-create your scale model to an actual model. Remember that each 1 cm square represents 1 inch on the birdhouse. Don't forget to cut a hole in the front of the house **before** you fold it into a three-dimensional house.

6. You can use available materials to decorate your house either before or after you fold it into its shape. Markers will work just fine.

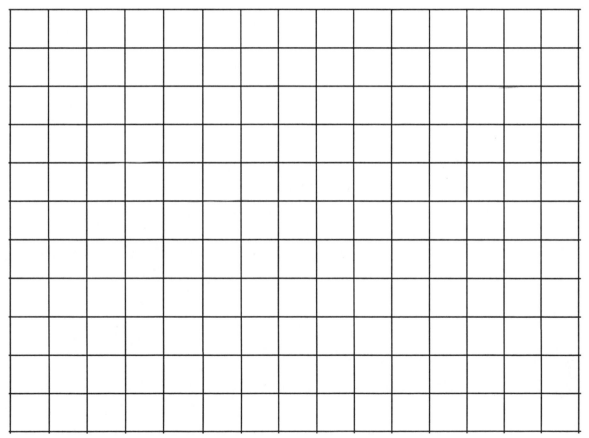

Birdhouse Nets

☞ Describe the shape(s) that make up the front and back of your birdhouse. What polygons are they?

_____

_____

☞ Describe the shape(s) that make up the sides of your birdhouse. What polygons are they?

_____

_____

☞ Describe the shape(s) that make up the roof of your birdhouse. What polygons are they?

_____

_____

☞ What formulas would you use to find the area of these shapes?

_____

_____

☞ What is the surface area of your birdhouse? Surface area is the sum of the areas of each of the surfaces (front, back, sides, and roof).

_____

☞ If you had to pay 5 cents for each square inch of surface area, how much would it cost you for paint for your birdhouse? Explain how you solved this problem.

_____

_____

Monthly Mortgage Payments

Areas of Study

Percentages, computation, rounding, reading employment ads

Concepts

Students will:

- use the newspaper to find a job with a listed salary
- compute their monthly income
- ascertain their acceptable monthly mortgage payment

Materials

- Monthly Mortgage Payments handout
- employment ads
- scissors
- glue
- calculator

Procedures

Students will search through the want ads in a local paper and find five jobs with a salary listed.

They will paste these ads on the handout in the space provided.

They will then calculate their monthly salary (to the nearest $10) for each job. Lending institutions require that a monthly mortgage payment not exceed 28 percent of the gross monthly salary. Students need to calculate 28 percent of each salary, rounding to the nearest dollar.

Assessment

1. Student product:
   - worksheet
2. Observation of students
3. Journal question(s):
   - Not all want ads list the salary. Why do you think employers would advertise a job opening and not be specific about salary?

Extension

- Research want ads and compare monthly income with education and/or experience.

Monthly Mortgage Payments

Banks and other lending institutions investigate people before loaning them money for a house. Banks check your salary, your work history, and your credit report to find if you pay your bills on time. Banks often require that your monthly mortgage payments not be more than 28 percent of your gross monthly income.

Using the want ads of a local newspaper, find five ads for employment that list a salary or wages. Determine the monthly income for the job. Then calculate the maximum monthly mortgage payment, using 28 percent of the gross monthly income. Round the monthly income to the nearest $10. Round the mortgage payment to the nearest dollar.

Paste the newspaper ads in spaces below.	Monthly Income	Maximum Mortgage Payment

Monthly Mortgage Payments and Interest Rates

Areas of Study

Reading charts and tables, computation, problem solving

Concepts

Students will:

- use Monthly Mortgage Payment Chart to determine monthly mortgage payments
- compare mortgage payments based on rates of interest
- calculate the total cost of a home
- solve an open-ended problem

Materials

- Monthly Mortgage Payment and Interest Rates handouts
- calculator

Procedures

Distribute Monthly Mortgage Payment Chart to each student and discuss what each of the columns and rows represents. Explain how the interest rate affects the monthly payment.

Allow students to work in pairs to answer questions on the handout. Question 4 has multiple solutions, and students should be given the opportunity to explain how they solved it.

Solution

1. $624
2. $20,000 + (30 × 12 × 790)

$20,000 + $284,400

$304,400

3. Both payments are $420. The payments are given to the nearest dollar. The payments calculated to the nearest cent would have different values.

4. $95,000 at 6.5%

$90,000 at 7%

$85,000 at 7.5%

$80,000 at 8%

$75,000 at 8.5%

$70,000 at 9.5%

$65,000 at 10.5%

Assessment

1. Student product:
   - completed worksheet
2. Observation of students
3. Journal question(s):
   (a) Explain why it is just as important to shop around for a good interest rate as it is to find your dream house.
   (b) If the purchase price of a house is $110,000 and the actual payments amount to $292,000, what is the percent of increase?

Extension

- Research the cost of homes in your area as well as rental rates. Compare the monthly cost of each.

Monthly Mortgage Payments and Interest Rates

Banks and other financial institutions charge different interest rates for mortgages. It is just as important to shop for the best interest rate as it is to shop for the best home. If you pay a 20 percent down payment for a house, you must borrow the remainder of the cost. Most home mortgages are for 30 years. The chart below lists the monthly mortgage payment you must pay. This is only the fee for the principal and interest. Taxes and insurance are not included.

Monthly Mortgage Payment Chart

Loan Amount	Interest Rates								
	6.5%	7%	7.5%	8%	8.5%	9%	9.5%	10%	10.5%
$ 20,000	$128	$133	$140	$147	$154	$161	$168	$176	$183
25,000	158	166	175	183	192	201	210	219	229
30,000	190	200	210	220	231	241	252	263	274
35,000	221	233	245	257	269	282	294	307	320
40,000	253	266	280	294	308	322	336	351	366
45,000	284	299	315	330	346	362	378	395	412
50,000	316	333	350	367	384	402	420	439	457
55,000	348	366	385	404	423	443	462	483	503
60,000	380	399	420	440	461	483	505	527	549
65,000	411	432	454	477	500	523	547	570	595
70,000	442	466	489	514	538	563	589	614	640
75,000	474	499	524	550	577	603	631	658	686
80,000	505	532	559	587	615	644	673	702	732
85,000	537	566	594	624	654	684	715	746	778
90,000	569	599	629	660	692	724	757	790	823
95,000	600	632	664	697	730	764	799	834	869
100,000	632	665	699	734	769	805	841	878	915
150,000	948	999	1050	1101	1152	1206	1260	1317	1371
200,000	1264	1330	1398	1468	1538	1610	1682	1756	1830
250,000	1580	2328	1748	1835	1993	2005	2103	2195	2288
300,000	1869	1995	2097	2202	2307	2415	2523	2634	2745
500,000	3160	3325	3495	3670	3845	4025	4205	4390	4575

Monthly Mortgage Payments and Interest Rates

Use the Monthly Mortgage Payment Chart to help you answer these questions.

1. What would your monthly payment be if you borrowed $85,000 at 8% interest?

   _____

2. If you made a $20,000 down payment and borrowed $90,000 at 10% interest you would pay $790 monthly for 30 years. How much did you actually pay for the house (one payment a month for 30 years)?

   _____

3. Compare the monthly payments on a $60,000 loan at 7.5% and a $50,000 loan at 9.5%. Why does this occur?

   _____

   _____

   _____

4. You can afford a monthly mortgage payment of $600. List your loan and interest rate possibilities.

   _____

   _____

   _____

   _____

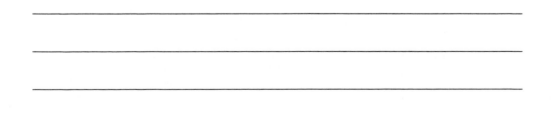

$25,000 Decorate-Your-Dream-Bedroom Contest

Areas of Study

Computation, scale drawing, problem solving, reading catalogs and newspaper ads

Concepts

Students will:

- use newspaper ads and catalogs to find items to redecorate their bedroom

- record purchases and itemize costs, keeping a running total

- draw a scale model of their bedroom, making sure that all items will fit in the room

- find a solution that comes as close to $25,000 as possible

Materials

- catalogs and advertising sections from news-papers

- $25,000 Decorate-Your-Dream-Bedroom Contest handouts

- calculator

- scissors

- glue

- markers or crayons

- ruler

Procedures

Give each student a copy of the project and discuss the rules of the contest. Go over the guidelines with students, emphasizing that the items must be tangible and available for purchase in your area. Start saving Sunday supplement sections and collecting gift catalogs well in advance of starting the project. Read through each rule and be sure that students understand the concept.

Once this is done, students can begin working on the project, using available reference materials. This project will take three days to complete, and students will need to do some of the work at home.

Assessment

1. Student products:
   - completed forms
   - scale drawing
2. Observation of students
3. Journal question(s):
   - How much more would your selections have cost if you had had to pay a 6-percent sales tax on each of the items?

Extensions

- Have students redecorate an entire house.

- Have students purchase new equipment and desks for the classroom, using $25,000.

Name _____ Date _____

$25,000 Decorate-Your-Dream-Bedroom Contest

Pretend you have won $25,000 tax free to finance a dream bedroom. You may purchase clothes to fill your closet. You must spend the $25,000 on tangible items. All the items must fit in a 25-foot by 25-foot bedroom.

You must follow these guidelines.

1. You may buy only one of each item: one bed, one lamp, one mirror, one exercise bike, one computer.

2. You must buy **at least 10 items** and **as close to 20 items** as possible.

3. You will ignore sales tax. The money is tax free.

4. Your goal is to spend **exactly** $25,000. Do not round. The $25,000 must be to the penny.

5. Allow lots of time to work on this project—$25,000 is a great deal of money to spend.

6. Your $25,000 Dream Bedroom Tally Sheet must be neat and organized. It should be accurate and should show effort. You may need more than one tally sheet.

7. A scale model of your dream bedroom showing the location of your purchases must be included with a clipping, drawing, or description of each item.

$25,000 Decorate-Your-Dream-Bedroom Contest

$25,000 Dream Bedroom Tally Sheet

Source (Where did you buy it? Who told you the price?)	Item Description	Item Total Dollars and Cents	Running Total Dollars and Cents
		Total for this page	

Name _____ Date _____

$25,000 Decorate-Your-Dream-Bedroom Contest

Scale Model: 25 Feet by 25 Feet

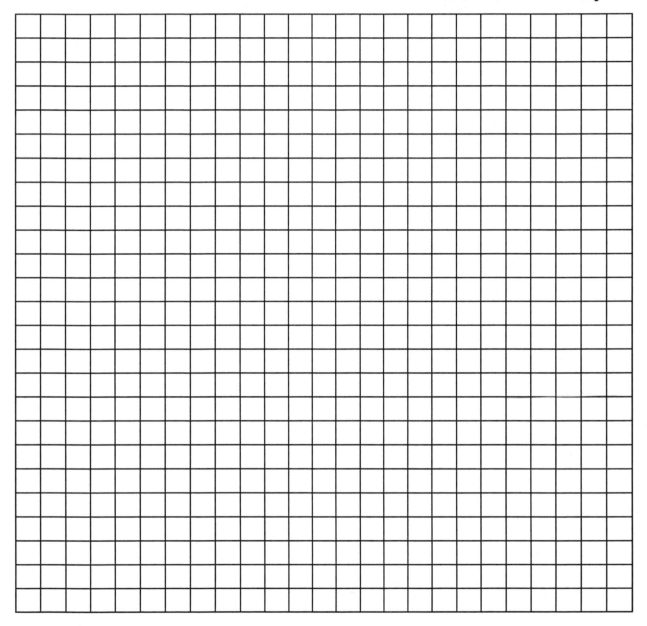

Math and Poetry

Procedures

One way to integrate mathematics with other subject areas is to include some interesting poetry assignments as part of the math lesson. The three poems included in *Integrating Math in the Real World: The Math of Homes and Other Buildings* are acrostic, cinquain, and haiku.

Acrostic poems have the letters of a word or phrase written vertically. The students are asked to make a poem from My House. Each line of the poem begins with the corresponding letter of the title. A word, a phrase, or a sentence can be written on the line. The line must refer to the title of the poem and cannot merely be a collection of words that begin with that letter.

Cinquain poems consist of five lines. The first line is one word. The second line is two words. The third line is three words. The fourth line is the author's personal reaction to the topic. The fifth line is a single-word synonym of the first line. Each of the words refers to the title (the first line).

Haiku is a three-line poem in which each line has a definite syllable count. Line one has 5 syllables, line two has 7 syllables, and line three has 5 syllables. The best haiku have a twist of fate and are illustrated. Space is provided for the illustration.

The following matrix is suggested as a means of assessing the poetry:

Criteria	4	3	2	1
How well did the student follow the guidelines for poetry type?				
How correct is the grammar?				
How creative is the poem?				
What is the overall quality of the project?				

Math and Poetry

Acrostic

The title of this poem is *My House*. The title is also written vertically so the first letter of each line corresponds with a letter of the title. Write a word, phrase, or sentence that refers to the title and is not just a selection of words that fit the letters.

My House

M

Y

H

O

U

S

E

Math and Poetry

Cinquain

A cinquain is a poem that has five lines. The first line is a single noun that is the subject of the poem. The second line consists of two adjectives that describe the noun. The third line has three words: at least one is a verb and another an adverb. The fourth line is personal and contains four words describing how the author feels about the subject. The last line is a single-word synonym for the first line.

Home

_____

_____   _____

_____   _____   _____

_____   _____   _____   _____

_____

Math and Poetry

Haiku

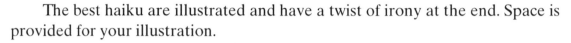

Haiku poems have a specific syllable count and just three lines.

Line one has 5 syllables.

Line two has 7 syllables.

Line three has 5 syllables.

The best haiku are illustrated and have a twist of irony at the end. Space is provided for your illustration.

My Favorite Building

_____ (five syllables)

_____ (seven syllables)

_____ (five syllables)

Buying a House You Can Afford

Areas of Study

Computation, percents, reading tables and charts, research skills, open-ended problem solving

Concepts

Students will:

- assume the identity of a person wishing to buy a house

- determine their potential yearly income from employment ads

- calculate the size of their house payment based on yearly income and down payment available

- use a table to determine how much money they can borrow to purchase the home

- use real estate ads to find their affordable dream home

- complete a display sheet containing their calculations and a picture of their dream home

Materials

- newspapers
- real estate home-buying guides (collected from local real estate agents)
- Financial Profile Card
- Buying a House You Can Afford handouts
- calculator
- glue stick
- scissors

Procedures

Make copies of financial profile cards (pages 62–64) and cut them into individual cards. Have students select one randomly from a bag or a hat.

Once students have their card, they will know the type of job for which they are qualified. Each student (or pair of students) should have a copy of the employment section from a local newspaper or other source. Using this, they can estimate their potential yearly income and their monthly income. Charts and tables have been supplied to help students compute the value of the house they can afford. The Monthly Mortgage Payment Chart (page 65) lists the payment per month, based on different interest rates. This payment may not exceed 28% of the monthly income if the student intends to get a loan from a bank. The Buying a House You Can Afford handouts require students to: (1) find a job and determine their potential income (the employment ad must be included in their final presentation), (2) calculate their monthly income and 28% of that, (3) determine how much money they can borrow, using the Monthly Mortgage Payment Chart, and (4) find a house for that amount of money.

Assessment

1. Student product:
   - completed handout

2. Observation of students

3. Journal question(s):

   (a) Explain why it is advisable to have the largest down payment you can afford.

   (b) Three families—the Smiths, the Blacks, and the Martins—all pay approximately $505 per month in mortgage payments. But the Smiths have a $55,000 home, the Blacks have a $65,000 home, and the Martins have an $80,000 home. How can you explain this?

   (c) Why do you think that banks stipulate that families should pay, at most, 28% of their monthly income for their mortgage? Why can't it be 50% or more?

(d) If you purchase a $100,000 house and your payments are $632 each month, how much will you actually pay for the house if you have a 30-year mortgage?

Extensions

- Students can research current mortgage rates at a variety of financial institutions to determine the best rates in your area.

- The class can keep track of actions of the Federal Reserve Board and interest rates.

Buying a House You Can Afford

Financial Profile Cards

Give all students in the class one of these cards. Each card has a very brief financial profile for use when purchasing a house.

Age: 18 Family: single Education and Training: high school graduate—no other training. Years of Experience: 2 years part-time at a fast-food restaurant Down Payment Available: $7,500	Age: 19 Family: engaged Education and Training: high school graduate and 1 year of computer training Years of Experience: none Down Payment Available: $5,000
Age: 20 Family: single Education and Training: junior college graduate—associate degree in restaurant management Years of Experience: 4 years part-time in local restaurant Down Payment Available: $9,000	Age: 21 Family: married Education and Training: completed an electrician apprentice program Years of Experience: 3 years Down Payment Available: $9,500
Age: 22 Family: single Education and Training: high school dropout—working on GED Years of Experience: 6 years as wait staff Down Payment Available: $3,000	Age: 23 Family: married with one child Education and Training: college degree with accounting major Years of Experience: 1 year as bookkeeper Down Payment Available: $10,000
Age: 24 Family: married with 1 child Education and Training: college degree in communications Years of Experience: 4 years part-time as a radio DJ Down Payment Available: $9,000	Age: 25 Family: single Education and Training: associate degree with a chef major Years of Experience: 5 years as an assistant chef Down Payment Available: $10,000

Buying a House You Can Afford

Financial Profile Cards

Age: 26 Family: engaged Education and Training: high school graduate with truck driver technical school Years of Experience: 2 years at a gas station and 5 years as a truck driver Down Payment Available: $15,000	Age: 27 Family: married with 2 children Education and Training: college degree in banking and finance Years of Experience: 2 years as head teller and 3 years as loan officer Down Payment Available: $20,000
Age: 28 Family: married with no children Education and Training: college degree in art and design Years of Experience: 6 years as designer in an ad agency Down Payment Available: $11,000	Age: 29 Family: divorced with 1 child Education and Training: GED and 1 year of computer training Years of Experience: 10 years in data entry Down Payment Available: $5,000
Age: 30 Family: married with 2 children Education and Training: master's in business administration Years of Experience: 5 years as finance officer of small computer firm Down Payment Available: $25,000	Age: 31 Family: married with 3 children Education and Training: completed apprentice program in plumbing Years of Experience: 8 years as an independent plumber Down Payment Available: $22,500
Age: 32 Family: married with 1 child Education and Training: college degree—major in nursing Years of Experience: 8 years Down Payment Available: $12,500	Age: 33 Family: single Education and Training: completed law school and passed bar exam Years of Experience: 9 years experience as a public defender Down Payment Available: $21,000
Age: 34 Family: married with 1 child Education and Training: teacher with a master's degree in education Years of Experience: 12 years Down Payment Available: $19,500	Age: 35 Family: married with no children Education and Training: completed medical school with a residency in pediatrics Years of Experience: 5 years Down Payment Available: $15,000

Buying a House You Can Afford

Financial Profile Cards

Age: 22 Family: single Education and Training: college degree in forestry Years of Experience: 4 years part-time as a summer ranger Down Payment Available: $4,000	Age: 23 Family: married with no children Education and Training: associate degree in landscaping Years of Experience: 5 years part-time (golf course) and 2 years full-time (own business) Down Payment Available: $10,000
Age: 24 Family: single Education and Training: high school Years of Experience: 6 years at large department store Down Payment Available: $5,500	Age: 25 Family: married no children Education and Training: college degree in engineering Years of Experience: 3 years Down Payment Available: $22,500
Age: 26 Family: married with 1 child Education and Training: college degree in journalism Years of Experience: 4 years at local paper Down Payment Available: $11,000	Age: 27 Family: divorced with 2 children Education and Training: college degree in foreign languages Years of Experience: 5 years as a translator for an international company Down Payment Available: $19,500
Age: 28 Family: married with no children Education and Training: high school Years of Experience: 10 years at a large copy service Down Payment Available: $9,000	Age: 39 Family: married with 2 children Education and Training: high school diploma and military training Years of Experience: retiring after 20 years in the Army Down Payment Available: $25,000
Age: 30 Family: married with 3 children Education and Training: high school and associate degree in structures Years of Experience: 10 years as foreman for construction company Down Payment Available: $12,500	Age: 31 Family: divorced with 1 child Education and Training: college degree in business and real estate course Years of Experience: 9 years as real estate agent Down Payment Available: $13,000

Buying a House You Can Afford

Monthly Mortgage Payment Chart

The rules of our lending institution are as follows: In order to qualify for a 6.5% mortgage rate, you need to make a down payment of 20% or more. To qualify for an 8.5% mortgage rate, you need a down payment of at least 10%. And if you can make a down payment of only 5% (the minimum), you will have to pay an interest rate of 10.5%. The dollar amounts under each percent reflect your monthly mortgage payments.

Loan Amount	Interest Rates		
	6.5%	8.5%	10.5%
$20,000	$128	$147	$183
25,000	158	192	229
30,000	190	231	274
35,000	221	269	320
40,000	253	308	366
45,000	284	346	412
50,000	316	384	457
55,000	348	423	503
60,000	380	461	549
65,000	411	500	595
70,000	442	538	640
75,000	474	577	686
80,000	505	615	732
85,000	537	654	778
90,000	569	692	823
95,000	600	730	869
100,000	632	769	915
150,000	948	1,152	1,371
190,000	1,201	1,461	1,738
200,000	1,264	1,538	1,830
250,000	1,580	1,993	2,288
300,000	1,869	2,307	2,745
500,000	3,160	3,845	4,475

Buying a House You Can Afford

```
┌─────────────────────────────────────┐
│                                     │
│                                     │
│                                     │
│      Paste your Financial Profile Card here      │
│                                     │
│                                     │
│                                     │
└─────────────────────────────────────┘
```

With my education and experience, I qualify for the following job: _____

_____

My potential yearly income is: _____

```
┌─────────────────────────────────────┐
│                                     │
│                                     │
│                                     │
│        Paste your job ad here        │
│                                     │
│                                     │
│                                     │
└─────────────────────────────────────┘
```

Use the following example to help with your calculations. Let's assume that you have $10,000 for a down payment. If you buy a $50,000 home, you have a 20 percent down payment, because $10,000 is 20 percent of $50,000. By putting down 20 percent, you qualify for a 6.5 percent loan, and so your mortgage payment will be $253 per month for a $40,000 loan. But if you buy a $100,000 home and you put down 10 percent of the price, your interest rate will be 8.5 percent, and your monthly payment will be $692 for a $90,000 loan. It is possible to put down only 5 percent; this will allow you to buy a house for $200,000, but then your interest rate will be 10.5 percent, and your monthly payment will be $1,738 for a $190,000 loan. You really do not have all these options available to you, however. Remember that the bank has restrictions and will lend you only a limited amount based on your income.

Buying a House You Can Afford

What if you choose to buy the $200,000 house? Your monthly payment cannot exceed 28 percent of your monthly income. To make a monthly mortgage payment of $1,738, you are required by the banks to have an income of at least $6,207.14 per month. $1,738 is 28 percent of $6,207.14. Make sure you look carefully at what the bank says you can afford to pay in monthly mortgage payments.

My monthly income will be: _____

My monthly mortgage payment can be (28% of my monthly income): _____

My Financial Profile Card shows that I have a down payment of: _____

Look at the Monthly Mortgage Payment Chart.

Your decision will need to be based upon: (1) how big your mortgage payment can be and (2) whether you have the down payment to qualify for this house.

My decision is to buy a house that costs _____ and to make a down payment of _____ %. I will be paying _____% interest on the loan, and my monthly payments will be _____.

Paste a picture of your affordable
dream house here.

Housing Opportunity Index

Areas of Study

Computation, percentages, reading charts and tables

Concepts

Students will:

- calculate the down payment needed to purchase a median-priced home in each city

- calculate the monthly mortgage payment based on the median income of the city's residents

Materials

- Housing Opportunity Index handouts
- calculator

Procedures

Review the meaning of median as a measure of central tendency. Discuss the housing index and differences by region and city. Students will calculate the down payment needed to buy the median-priced home and the maximum monthly mortgage payment.

Solutions

City	Housing Opportunity Index (HOI)	Median Home Selling Price	20% Down Payment	Median Yearly Income	Maximum Monthly Mortgage Payment
Akron	68.8	$95,000	$19,000	$44,300	$1034
Boston	70.7	146,000	29,200	59,600	1391
Chicago	64.0	140,000	28,000	55,800	1302
Honolulu	40.9	226,000	45,200	57,900	1351
Kokomo, IN	89.5	85,000	17,000	46,900	1094
Los Angeles	50.2	158,000	31,600	47,800	1115
San Jose, CA	39.4	260,000	52,000	70,200	1638
Ocala, FL	79.6	69,000	13,800	33,300	777
Kansas City, MO	84.1	91,000	18,200	50,200	1171
Washington, D.C.	72.8	160,000	32,000	70,300	1640
U.S. Average	66.5	120,000	24,000	43,500	1015

Assessment

1. Student product:
   - completed handout
2. Observation of students
3. Journal questions:
   (a) Why would communities brag about a high Housing Opportunity Index rating?
   (b) Why is median used rather than mean or mode?

Extension

- Research the Housing Opportunity Index for your local community and compare it to the national average.

Housing Opportunity Index

The Housing Opportunity Index (HOI) measures the affordability of homes by metropolitan regions. The index first determines the median income and median selling price of homes. The HOI is the percentage of people earning the median income who can afford to buy a home at the median price.

In 1997, Kokomo, Indiana, was the nation's most affordable housing market, according to the HOI. The median family income in Kokomo was $46,900. The median price of homes sold was $85,000. The HOI of 89.5 means that 89.5 percent of the families in Kokomo earning the median income could buy a house for the median price. In contrast, San Francisco has the nation's least affordable homes. The median income is $64,000 and the median home price is $288,000. The HOI is 23, meaning that only 23 percent of the people earning the median income can afford a home selling at the median price.

Most people need to borrow money to buy a house. The bank or lending institution places restrictions on the amount of money you can borrow. Most banks require a down payment of at least 20 percent, and your maximum month mortgage payment cannot be more than 28 percent of your monthly income.

The chart on page 70 lists the city, HOI, median home price, and median income. Calculate the 20 percent down payment needed and the maximum monthly mortgage payment.

(continued)

Housing
Opportunity Index

City	HOI	Median Home Selling Price	20% Down Payment	Median Yearly Income	Maximum Monthly Mortgage Payment
Akron	68.8	$95,000		$44,300	
Boston	70.7	146,000		59,600	
Chicago	64.0	140,000		55,800	
Honolulu	40.9	226,000		57,900	
Kokomo, IN	89.5	85,000		46,900	
Los Angeles	50.2	158,000		47,800	
San Jose, CA	39.4	260,000		70,200	
Ocala, FL	79.6	69,000		33,300	
Kansas City, MO	84.1	91,000		50,200	
Washington, D.C.	72.8	160,000		70,300	
U.S. Average	66.5	120,000		43,500	

Web Sites, Books, and Pamphlets

http://www.fanniemae.com/index.html
   the Fannie Mae Foundation, the largest source of home mortgages, and its links have a tremendous amount of information on mortgages, financing a home, and educational materials.

http://www.nahb.com/pg2.html
   contains Housing Opportunity Index and housing market information.

http://www.cypressmtg.com/faq.html
   a list of frequently asked questions concerning mortgages and financial institutions.

http://www.bayhouse.com/mortgage.html
   contains a calculator to help with monthly mortgage payments. Enter interest rate, amount of down payment, and home price. Instructions on creating spreadsheets to calculate monthly mortgage payments.

http://www.homefair.com/homefair/mortwork.html
   contains a worksheet for prospective home owners that helps decide the affordable amount of down payment and monthly payments.

Brooks, Hugh. (1992). *Building & Construction Resource Directory.* New Port Beach, CA: HBA Books. List of names, addresses, and phone numbers of major building trade associations, lists of colleges and universities with construction-related programs, names of contractors, engineers, and architects, and reference books by subject matter.

National Trust for Historic Preservation. (1985). *Built in the USA.* Preservation Press. Facts on U.S. buildings from airports to zoos. Historical look at many U.S. buildings with lots of photos and Americana.

Olsen, Michele Rojek, and Olsen, Gary L. *Archi-Teacher: A Guide to Architecture in the Schools.* Champaign, IL: Educational Concepts Group. A guide for elementary teachers to present architecture to students by using activities in math, art, history, and science.

Opening the Door to a Home of Your Own. Fannie Mae Foundation, 4000 Wisconsin Avenue NW, Washington, D.C. 20016-2800. A free guide to getting your first mortgage. Charts and explanation of the process of qualifying for a mortgage.

Robson, Pam. (1995). *SHAPES Structures and Material.* New York: Shooting Star Press. Student projects ideas with scientific reasons. Math and science connections.

Salvadori, Mario. (1990). *The Art of Construction.* Chicago: Chicago Review Press. Explains how tents, houses, stadiums, and bridges are built. Student lessons show how to build structures and models using materials found around the house.

Stephen, Taylor. (1988). *A Place of Your Own Making; How to Build a One-Room Cabin, Studio, Shack, or Shed.* New York: Henry Holt and Company. A guide to basic building techniques with illustrations and vocabulary.

Williams, Gene. (1990). *Be Your Own Architect.* Blue Ridge Summit, PA: Tab Books. Sample plans, basic construction information and techniques.

Wylde, Margaret; Baron-Robbins, Adrian; and Clark, Sam. (1994). *Building for a Lifetime: The Design and Construction of Fully Accessible Homes.* Taunton Press, Inc. Source of floor plans and designs for accessibility, lists of resources, and adaptation requirements.

Share Your Bright Ideas with Us!

We want to hear from you! Your valuable comments and suggestions will help us meet your current and future classroom needs.

Your name_____Date_____

School name_____Phone_____

School address_____

Grade level taught_____Subject area(s) taught_____Average class size_____

Where did you purchase this publication?_____

Was your salesperson knowledgeable about this product?   Yes_____   No_____

What monies were used to purchase this product?

___School supplemental budget   ___Federal/state funding   ___Personal

Please "grade" this Walch publication according to the following criteria:

Quality of service you received when purchasing ....A   B   C   D   F
Ease of use...A   B   C   D   F
Quality of content...A   B   C   D   F
Page layout ....A   B   C   D   F
Organization of material ....A   B   C   D   F
Suitability for grade level ....A   B   C   D   F
Instructional value..A   B   C   D   F

COMMENTS:_____

_____

What specific supplemental materials would help you meet your current—or future—instructional needs?

_____

Have you used other Walch publications? If so, which ones?_____

May we use your comments in upcoming communications?   ___Yes   ___No

Please **FAX** this completed form to **207-772-3105**, or mail it to:

Product Development, J.Weston Walch, Publisher, P.O. Box 658, Portland, ME 04104-0658

We will send you a **FREE GIFT** as our way of thanking you for your feedback. **THANK YOU!**